Tangled *Yet* Coiled

Tangled, Yet Coiled
by Tom Burgess

© Tom Burgess

ISBN: 9781912092192

First published in 2021
by Arkbound Foundation (Publishers)

Cover image: PM 28 by Carol Brown Goldberg © *Private Collection*

No part of this publication may be reproduced, stored in a retrieval system, or transmitted, in any form or by any means without the prior permission of the publisher, nor be otherwise circulated in any form of binding or cover other than that in which it is published and without a similar condition being imposed on the subsequent purchaser.

Arkbound is a social enterprise that aims to promote social inclusion, community development and artistic talent. It sponsors publications by disadvantaged authors and covers issues that engage wider social concerns. Arkbound fully embraces sustainability and environmental protection. It endeavours to use material that is renewable, recyclable or sourced from sustainable forest.

Arkbound, Backfields House
Upper York Street
Bristol, BS1 8QJ

www.arkbound.com

Tangled *Yet* Coiled

Foreword

By Tracey West

CEO and Co-Founder of
The Word Forest Organisation

I've been a writer and broadcaster on sustainable living since my 30s. I proudly launched The Book of Rubbish Ideas on ways to reduce landfill waste and, later that decade, by some strange twist of gut-wrenching fate, I became a stand-up performance poet on divorce and the environment. Not usually in the same poem, I might add!

I turned 55 on 21st March 2021, which is also the International Day of Forests. What are the chances of a co-founder of a small international reforestation charity being born on that day? What's interesting about the International Day of Forests is that it's only 9 years old and I'd been hopping around the planet for over 40 years before it existed.

It's a wonderful sign of the times that I'm not entirely

alone in wanting to celebrate my over half-century in a non-consumerist way. Many, like me, agree they're also happier living with less. In a world positively drowning beneath adverts espousing how much better our lives would be if we had a (...fill this blank space with a variety of widgets or gadgets...) I'm encouraged by that.

Materially, I need nothing. My clothes are all proudly procured from local charity shops, I have no personal requirement for tech or gadgetry, my darling husband and I have a second-hand EV which gets us where we need to go and we have a beautiful rented roof over our heads in a space we love.

When you've spent time with the amazing people of rural Kenya, who typically live in simple mud huts and have a handful of possessions to their names (frequently a couple of pots for cooking, some 10 and 20 litre water containers, a few items of clothing, a bowl and a wheelbarrow if they're incredibly lucky) you quickly realise that a minimalist existence in the UK comprises far more 'stuff'. Whilst on our Monitoring and Evaluation visits to our projects in Kenya, I developed an even greater understanding of the term 'first world problems' and a far clearer definition of the words 'want' and 'need'.

Maslow's hierarchy of needs presents the various levels of human requirements as: 1) physiological; 2) safety; 3) belonging and love; 4) social needs or esteem, and 5) self-actualisation. The final one is recognised as the level where we achieve our full potential and explore creative abilities. I'd say without a bountiful and regular dose of level 5 - with stories, big art and poetry - I'm not entirely whole.

When Tom first mentioned 'Tangled Yet Coiled' my senses went fizzy. The thought of sitting peacefully under a tree, meandering slowly through a lovingly created collection of poetry that embodies some of my favourite things, was thrilling! All of a sudden, here it is.

Tom's work is standing silently in the wings, dusting itself down, waiting to take centre stage and preparing to enter your eyes, rise to your frontal creative cortex and fill your soul with utterings of simplicity, beauty, power, majesty, fragility and optimism.

Together, we are living, breathing creatures dotted around in all sorts of places, all around the world. We breathe together as one.

Preface

Who understands the wind?

Who can listen to the trees?

Who can decipher the messages of fossilised rain drops?

This is a collection of poems with the unifying themes of nature, ecological breakdown and those wilderness experiences that vault the ego.

I have been on a journey of mental instability and absolutely vital to my path of recovery has been the role of nature in offering connection; a sense of self and place. For that I am deeply grateful and I acknowledge how lucky I am to have been able to access these places where nature gives so much. By way of introduction, I include an extract from my diary:-

'Goodness spreads and is emulated. Music gets played in different contrasting places, quality and joy leapfrog division and barriers. There is so much confusion in the world, there is so much confusion within me. We are stretched by different perspectives. All informed by different sources, never before have I wanted so badly for the truth to rise up. Whose truth, that is just it, I do not wish to be the judge. I want a Truth that emerges uncontested and unifying.

There is Climate breakdown truth, a sobering consensus that provokes action. I run out of ideas on what to do, I get bored, I dither and don't commit to anything. Frozen in time, I am stuck. I am complacent. I am not a committed rebel of XR - why not? I must forgive myself for that, I have joined in, but I know my floodgates remain barricaded. How should I adjust my life appropriately? There is a huge chasm between what I hope for and the reality. I want to produce something of persuasive beauty, as though the world will be sorted with words not deeds. Though first comes story - not these thoughts but something better, a narrative we can get behind. One is emerging; the scale of direct action and protest is a signal that a brewing story decades' old is now coming to the boil. A zeitgeist wave is being ridden, but is there time? What of me, standing back again, a witness not rushing to ride the same wave and add to its velocity?

What happened to my dreams? I am too scared to have them, or too lazy. Or has immanent existential dread rendered only collective dreams worth pursuing? Yet, my dreams were always orientated around community. My core energy has dwindled and rotted. There is so much beauty in this world and unimaginable pain; both are beyond being contained by me; both outstrip my cupped hands. What of those young people who are being inducted early into the injustice of the world, its corruption and brokenness? Perhaps their action opens them up to a growth and beauty I can only imagine.'

I write these poems to initiate change in myself; I share them as an act of solidarity.

Contents

Foreword	I
Preface	V
To The Echelons of Power	1
Chatter, Speech, Screams	3
Slate	4
Revival	7
Wounded Healers	9
The line is Moving	10
Natural Shift	12
Incantation	14
Future Hedgerows	16
Tourist in Mumbai	19
Restless	20
A Brief Interlude	22
Ravelling	24
The Underwater Sound of Heavy Hearts	27

Looking for Strength	30
Dissolving Walks of Ecstasy	32
Moon in Orbit	34
Comfort and Joy	37
Wonder Close	38
Standing Rock	40
On These Things and More	44
Occupy Plastic	48
I want to find my rebel heart	50
About the Author	56

To The Echelons of Power

To the echelons of power, wherever you are
It is not too late to turn, please
play a different role,
another world is waiting to manifest.

Won't you pacify the grotesque hubris which has you
funding private escape pods
destined for far off planets?
What do you have to offer the mystery of space?
Except a profound disconnection with your origin and kin.
The unfolding caverns of majesty require an imagination,
one that goes beyond your sterile lunges for survival.
You vampires of the commonwealth
do not build bunkers out of fear and petrol
give to regeneration!
Stop casting the most vulnerable as fodder

may the word collateral choke in your throats.

What remains still extends grace towards you,
still the trees suffocated by indifference give oxygen,
and the blessing of a raindrop on your skin is a whisper.
Do not only hear a warning of scarcity and acid,
in that whisper hear hope too
and turn -
join in and work for regeneration!

Chatter, Speech, Screams

Anthropocene, we know you will not pause
Gears grind gung-ho, macho ideas win
for now, drowning rains we debate the cause
We callous conquerors turn from dark skin

Elemental fury has me shaking
Land ice is sweating to kill in cold blood
The seas surge god like, human hearts aching
Flood after flood after flood after flood

The angry planet holds some special good
You can hear it at the mouth of a brook
Each tree has a tongue speaking for the wood
The loose jaw of nature an endless book
We have to realign, do something and weep
Change is here, we are all called to leap.

Slate

When the rains come streams furrow the brow of the fell

Slate slabs claw raise and crag

Metamorphose valley written on fluid rock

The layered wilds of cyan stone groan, with age

Scratched at land invites me

To decipher her scrawl

A story born of fire

Volcanic violence ushered in ash

Squeezingitselfbreathless deep annunder hard rock reached for air

Harsh forms waiting to become

The fox's borran and outpost to the savage rook

Mined at Honister, slate lined faces smashed at ancient crease

Struck cold cleavage shedding layers with skin

Hard men peeling back the passage of time

Millennia mulling over change

Suddenly static strata

Skimmed by laal bairns across the dub

Glistening wet exposed

Layer

Slowly something so cold

Comes to mean sanctuary

Homestead slate

At yam by the hearth annunder slate

Green grain stacked against rain

Drip grey roofs clad in moss

Layer

Family heat at kitchen table going nowhere

Layer

Offspring flake away your youth crumbled

Layer

Every sight a joy each breath bright

Layer

Spider's web holds weary walls

And between the chinks

Stiff wisdom knits fractured memories

Lays still

Faces to the ground, The Gravestone Gurn

Slate the thread in life now dead

Revival

We are the climatic pivot of creation

if only we would join in

We do not need to be the heroes in creation

only join in

This blue dot hangs in the balance

Shivering for perpetually new forms of harmony

On the other side waits chaos of a dismal kind

We teeter on the brink of diminishing life

Seduced by homogeny

We are pointlessly surrendered to entropy

The natural home of stalemate

What we do matters

If we are not awake we distort the weaving of all things

We are substance in the scale of change

Choosing life means becoming a gift

Anticipating the new to allow the flow through

Together we matter

Our magnetic cross channels elevate a third way

awe, pulsing and synergy making movement possible

Together we widen the range of love.

Wounded Healers

Humanity is in the business of brokenness

Busy transmitting our hurt to the trees

This industrious despair takes the form of carbon

Trees transform our pain into oxygen

Rather than pass on the damage

They honour the free exchange of perpetual growth

It is a gesture of a blueprint

A smudge that hints at an interconnected dance

The original wounded healers

Hold wholeness

Centre themselves

And breathe

Giving life.

The line is Moving

Sat between grasping chaos and great potential

One side despair the other a brightening light

Indifference the broker for a careless evil

The line is thin

the line is brittle

The line is not straight

For one moment in time

A complex whole transmuted into a binary choice

The line is precious

the line is moving

To the magicians who desire a better world

Who create and toil

Who show up for it

Knock on the door for it

See dots

and connect them

Sweating to fuse them together

in the uncertainty of forging heat

Trusting in a larger frame

To all those who think beyond themselves

The line is precious

The line is moving

Natural Shift

A heart of longing

Body straining for its centre

In an airport sat separate

Willing my natural self to birth

from concrete

From synthetic light and tin music

The thunder of suitcases rolls on

I focus on the tingle of breath on my lip

and the feel of a thumb stroking forefinger.

These lungs, these many lungs like a forest

Humans huddled together in their inner worlds

Each life a cocktail of stories half told

and emerging

Where is nature's place in this story now

We are it

We belong in a family of beings and life

That no longer finds room or breath

Here

Solo knots tied intricately, as a net

That still shimmers in the right light

I remember meeting those three trees

Nature's ministry to me

A homecoming

That seat by the river when my edges disappeared

When life infused me and cartwheeled in and out and

around my centre, until

I was no longer the centre

In a constant orbit dance with everything else

The steep trees, dip green moss, bright boulders and ringing stream

All dancing

Particle to particle and eye to eye.

Incantation

My sphere is here, only now

single flow deliberate

true in this moment

Incantation

My sphere is here, only now

flutter of feathers

echo of a full chest

wind swaying branches

wind swaying leaves
The many layered sound of wind
A kind of silence

Again a sphere is drawn.

Future Hedgerows

Life is now charging down lanes in a personal train

Flanked by ancient monuments where plenty happened pre-parish life

Oblivious to boundaries huddle

The secret senate of birds still alive with sex buzz and industrial vigour

Life now throwing fumes in rash assault

The in-between places insecure

Birds dart at my wheels in protest

Ripples of the things race the windshield

Swerving hysterics wave us on down black knife tar

Such fanfare flurry within the bank awakens a Self That Soars

Fly by Hornbeam, Holly, Holm Oak and Hawthorn, race past the H's

Hazel replaced in the dictionary by futile acts of immortality

Red Backed Shrike a mere mortal that once forgotten can't be googled

Start with the art word sounds that sing
Starling, Wren, Song Thrush

Now the ideas we combine meet around wires grip

Soon a Nut Hatch will be in the palm of a child as they
gaze at the embers of another rainbow ended

Pressing at buttons whilst knowing little

The nameless animals have turned it up

To compete with the aching birth pangs of the city

They shout a shrill timorous song behind the
tweeting of pop sensation X

Now that the butterfly moves too slow for the fast lane the
whimsical lunges of the Romantics fade

We need a new urban doggerel that still shivers at words
like sunlight, leaves and dew

That sculpts in clutched pirouette with all that is
natural and true

Lyrics that dance like crazy with the pouring wonder
of this world within worlds

We don't need to be the climatic pivot of creation
only join in

Somehow content in the vast anonymity of slow time

I want to see hedges on sky scrapers

Let us redream the better rows of our ancestors

Redeem rootless prodigal landscapes all the way up

Oh, to bump into a Linnet or two on the way
to a conference on the 15th floor

Those grass-green heroes of older poems
could become associate members

Off to a problem-solving meeting met by the chirp
whistle of a new day

A bead sprinkle of light animated by bounteous rain drops

Lolling off leaves down glass windowed offices

Everything the right PH, might make us kinder

Like clean air to lungs, might make us bright

For the cartwheeling star that we call Sun

A quickening light, shines beyond this Tender country.

Tourist in Mumbai

Written in Abhanga - A Marathi poetry form

People perpetual

Incessant industry

Even this ancient tree

Cleaning the air

A few old trees remain

Gnarled cords push through concrete

The steady step of feet

Natural too

I touch wooden sinew

The pollution just one

Issue under the sun

Begin somewhere.

Restless

When we came to allocate the fields

Only those trees that fitted into our small horizons stayed

The survivors became nature's arbiters

Straddling the newly pastoral and a mythic world now gone

Traces left in gnarled bark and the whisper of leaves

Their lives prolonged by pollarding hand after hand after hand

Somewhere down the line hands were washed

Our connection further strained

Hedgerows no longer tended did not stay tame

Giving themselves in decay to the earth

Making fertile the soil, ground primed

the veterans remain

As wildness begins to reign we concrete over

Then desire a lawn

So buy fertiliser

Ship nourishment in from Russia, the US or Israel

Ever shaping the land

Restlessness the only story.

A Brief Interlude

At points, my own 'psychotic breaks' mirrored the widening fractures of this world. In between one of those phases I wrote the following in a diary. It emerged directly from an extended delusion too complex, fractured and kaleidoscope to describe clearly.

> *'Somehow, unwittingly as a result of my hubris and naivety, I am caught up in the end of the world. I have cast myself as super significant, contrary to all the evidence.*
>
> *A list of things that I hope are not my fault but I have a terrible sense they might be:*
>
> 1. *The universe is expanding, causing all that scary space*
> 2. *Division and the rise of the far right*
> 3. *Mass extinction*
> 4. *Freak weather events, infernal fires, dying sea algae, barren treeless wastelands*
> 5. *The confusion and chaos of fake news feeds and glamour of voices from multiple sides*

> 6. *Displacement of people from the terrors of war and climate breakdown*
> 7. *Food scarcity*
> 8. *Quiet loneliness and isolation*
> 9. *The universe is expanding, causing all that scary space*
>
> *I get stuck, I get indulgent, I get dramatic. I worry I am hurting those I love in inconceivable mysterious ways through my inability to transcend my egotistic grabs, groundwater fear and desire for control. For the sake of goodness, reach out and look beyond yourself.'*

Then amongst the same scrawls is the beginnings of a poem, which I wanted to preserve in its first form and that I am convinced belongs in this collection as it is. If only to underline and empathise with the enormous pressure, we can all put ourselves under and the strain we can take on if we do not meet ourselves with compassion.

Ravelling

Living in an ever tumbling

Swirling

universe

As it pours

itself out

And out, out

A huge

contorted feud rages

You realise

you contain the planet

reduced

down each rung of an ever more interconnected chain

Everything

you do, think even, obscures and contorts a better purpose

And, its

folding in on itself, the universe

Somehow

traveling from behind and in the future simultaneously

Tribes and

family's chasing their way through the medley
of DNA combinations

Friends and

all disparate connections are somehow fused

I find I am

the planet

Then the

planet in regression to something cold

Next I am a

silk like mist ever expanding racing outwards

As we fumble for solutions and adaptations (which will be a combination of structural and economic shifts, technological advance, lifestyle reorientation and conservation) let us remind ourselves we cannot solve these global environmental issues on our own and we are not to blame for them personally. First release, grieve, share and then take action.

The Underwater Sound of Heavy Hearts

Last night I was visited by Killer Whales

The thing is I never usually dream of other animals

We were on a kind of theme park glacier

The Whales grew increasingly impatient of the hoards

People who interpreted their presence as a sign

evidence that the human project was blessed

The scientist, she told me that the Whales
want to be acknowledged

That they wait vigorously to pierce our hubris

so they can broadcast their soundscapes of truth

Visceral messages which would melt our hearts
to water our brains

And dissolve our muddled certainty

These creatures are restless to repurpose confusion

They do not play with our fear, they seek to channel it

We have gathered tears from their tear-ducts for decades

Yet still ignore their grief

A groaning churning world preoccupies us

We dally with positive spins

Thin, cosmetic delusions

The longing for a panacea twists us up

corkscrewing our heads deeper into the sand

Swerving past the gritty grains of truth

As we dash around for cinematic moments of hope

And cast ourselves as saviour to the play things.

Play things they are not

They want to be met

They have been tapping out a message

Let's start there

Swimming through the broken ice of reality

Their heavy hearts are more than facts

Their swirling every day is not friendly

The life all around us is kinetic in its disbelief

Insisting on relatedness

Calling to our wild selves to awaken

And commune with the textured anguish of the deep sea

Their urgency will soon be ours

Discomfort crashing in with intuitive waves of knowledge

Intrusive is the truth

That time is up

Killer Whales burst in to my consciousness

Wanting to be acknowledged

Some humans have been diligently listening

I woke up from a dream to write this.

Looking for Strength

Beads of rain race

down electric wires

Fanned out from pylon, tracks form for waters rush

The precious glimmer of each dash

charged with suspense

Without warning they plummet

Gravity's kiss irresistible

I watch in a window of endless time

On multiple lines new translucent capsules emerge

Poised for their tight rope show

hesitant at first

Then in sheer delight at the thrill of life

They run in parallel lines

Clinging on

Gathering drops of water until the moment they fall

When at their fullest

Ripe wet jewels find my forehead

A blessing

Nature's gift.

Dissolving Walks of Ecstasy

What remains is a voice that was always there.

A nothing that needs not

a filling or stilted explanation.

Less thinking, mind sinking into a wider field of being.

A silenced sentience this lovely loosening

All is hungry

here

Yet woods beckon

The track winds to its end.

Sun on skin, back on

Back gone.

No path now,

just the quite tread of a new slower rhythm.

Desire changes its intensity, here drink

Anxiety stilled, no rush to remain quenched

No straining to hear the only sound that is.

Drenched in oneness

immerse into tune green and sunlight

the dance of dust pulses, sways with heat

filling lungs and senses.

Each step a deepening, bodily tingling of peace

and pleasure now entwined, indistinguishable

flung with time on air

Dissolving walks of ecstasy

Fling into surrender.

Moon in Orbit

Staring at the full moon
The last of a decade
Grateful eyes on a steady truth

It is the night before the election

Under the patient pearl of the moon
I am pacing and shaking
I have all sorts in me
All kinds of gravity

Hoping for the conviction to hope
The bravery to hope for something
I hope I have it
What a privilege this vague sentiment is
I have not done enough for those

That consider hope a luxury or a necessity

Instead of a plaything

Holding everyone I have ever known

somehow in time and space

Every moment of eye contact

Every conversation

Every stumbled exchange

Under the milk of the moon

The sum of every person's kindness

culminates and converges in unity

In moments of compassion and clarity

Scared that all this is just words

Next the cold practical considerations

Of getting through this life

Next to the bitter confusions of conflicting facts

And the anonymity of the non-gaze

Of that puckered lunar rock in the endless sky

So much pain and suffering in the world

Now is there more or is it less than before

Is that even the question

More people

More potential

More strain

We are a strain of something more

something larger

An effervescent

heart in a body responding to the moon

something vital is trying to gain traction.

Comfort and Joy

Let us

Knead love to the edges of our existence

Together become beings stretched

Striving for configurations of unity

A core pulsating to the outer limits

Where nothing rests

The abiding marrow still rich in nutrients

Origin of perpetual life

Love hewn from love then bonded new

Let us

live as though it were true

Rejoice in the luminous now

Rest in a kingdom of the eternal present

Amongst this new spaciousness

Find comfort in that elongated hum

Wonder Close

This track has led to desire's end

Rhythm has flung folds of emotion

to the floor

Tears clung to trees

in a huddle of friendship

Running back

arms outstretched

lips loose on a double truth

Body gently

speaking

Held in a sensuous net

Air blows through the

exposed pathways

of the brain

Trails once trodden

are being reclaimed

Hidden passages no longer overgrown

knitted new

A glorious tangle

To all things configured

To all reunited

Found just sitting

Like a shiver though an octave up

A resting place without shelter

Boundaries now contours

Appetite a silence

Thirst quenched then drained

then swimming

The cosmic nest never still but waiting

Standing Rock

These are not victims

These are rocks

These are water

Serenity's daughter combined with fires fury

This is a jury of beating hearts

Their same pulse permeates everything

for we are mud, all of us

nothing but mud

equally

As they school us in real learning

Everything that is earth Is Us

Not, an inanimate word for burning

These are not victims

They are power

Speak to aggressors with humility

You who stand there in the armour of the so-called free

To protect a bourgeoisie economy

Sioux extend love to you

See you as them

True love drowns fear

Transcends selfhood

Makes individuality look small.

So Sioux find you as shadows and not remotely scary

These are not victims

They invite you to stand with them like rock

As they fearlessly champion sanctity

They have waited

Centuries of oppression to be fully seen

And we all need them not the other way round

The main narrative in world history is indelibly at Standing Rock

A story that cries out incredibly for redemption

A ballad broadcast from the very pores of this planet

What the west lost in all its getting

Is a debt that we will all pay

Even the stones cry out.

The earth is not a victim

She invites you to join a resurgence of power

To fight for justice is what collective love looks like

Longing for mindful action the earth channels the mud

Perennial wisdom does rise

Status quo is disconnected and blind

The Native Youth's pounding feet get louder

The growing beat taps out a message

Exploitation will never 'trump' integration

nor stewardship, nor oneness

True expansion is towards greater unity

Line your pockets temporarily

Only

For you are not separate

and growth not exponential.

GDP don't start from free,

Inevitably it is built on reality

not infinity

So strive for harmony

for equilibrium

and back off.

On These Things and More

When all about you is tornado

And despair

Think on the unrelenting scale of space

And the persistence of a seed

Think on those eyes shut melodic moments

On limbs everywhere responding to rhythm

Think on those who share what they have

And when more room is made at the table

Count the ways we look beyond ourselves

When all about you is indifference

And cruelty

Dwell on ravished land restored

And moments of play and creation

Dwell on the glimmering touch of rain

And the vulnerability of honest communication

Live in the unfurling flourish of a fern

In that spiral eddy of a lazy river

Then measure the scale of our collective imaginations

When all about you is shaking

And sorrow

Keep pace with the surging voices for fairness

And the flight path of a swift

Keep pace with roads blocked in the name of peace

And every night spent under an open sky

On taking a stand and being stood by

See the shape of smoke at fireside cooking

Weigh the immediate joy of gratuitous giving

When all about you is yawning

And broken

Meditate on our distance from the stars

And your footprint on the sand

On the expediency of a feather

Hear stillness and singing and laughter

Meditate on first smiles and last smiles

And a gaze returned

On lifting up another's voice, on pause, on listening

Sense the pleasure in logic and the art of intuition

Yes, ruminate on justice, purity and goodness

Know the quiet truth of your changing feelings

Know that people are taking risks for beauty.

Occupy Plastic

He sits lotus like on concrete floor

Skin radiant with his smile

No food has passed his lips for sixteen days

Today is his birthday

This man named Sid takes only

the nourishment of his ideals

On the day of his birth

He pulses with the earth

Beautifully tethered to its worth

Not like that which he resists

A throw away culture still persists

Though in the light of his eyes it dwindles

That which is pulled from the bowels of the earth

To make a cameo appearance in the lives of humans

Then wretched and deformed it clings

to the worlds throat

Squeezing the life out of everything

People like Sid ask us to take note

To join in with our bodies and fertile minds

A natural intermingling for change.

I want to find my rebel heart

I am not laden in tattoos
Heavily bearded with a gentle swagger
I am not chained to an oil tanker
Locked on, arm tube my sleeve
With a heart of conviction and love

I am not but know not why

I want to find my rebel heart
I know it is there
I have felt it kicking, groaning
Weeping

This inner bird
Wings beating fast
Flying alone

Just about staying up in the storm

Waiting to be integrated

To come out of the shadows and grow strong

Can I channel my child heart?

That wildness, that curiosity

That lack of shame

Trust alive

Playing outside until it was dark

Scented air turned cool

Breathing deeply

Grass now damp under barefoot

Insistent that this can't end

Frantic facing the night

I remember climbing swaying trees

Swinging off the low hanging branches

Into waist high grass,

I'd stay crouching, held, still for a moment

The sound of crickets raucous and electronic

Long days, daisy chains and butter glow

Itchy scratches feet to knee

Sudden rain flicking up dust

Clothes on skin wet and grinning

Present as the waves

Paddling in the sea

Sand squeezed between toes

exploring rock pools

Waiting for their mysteries to be revealed

Poised, focused, expectant

These my first meditations

Such memories a profound privilege

How do I transfigure them into a live passion for rebellion?

Join with those brave and lonely souls who have been
sounding the alarm for decades

Such courage

Non-violent civil disobedience

Not utopian, limp or simplistic

A strong creative response to evil and indifference

My experiences are partial and soft

I know not of city childhood or a harsher natural world

Of scarcity and ferocious temperament

Right now as I meander with words defiant defenders

are being killed

Standing for their futures and the voices of their past,

the origins of myth

Tethered to legends, traditions laced with the roots of trees, connected

Standing for all, the host of abundant life that still remains

The Amazon both a picture book reality
and the air we breathe

Forest once grew where there is now a desert

A burning truth

Species that cannot be reanimated

Vaults of ice breached

Coral bleached

Top soil empty

Lack and unrest already conducting chaos across the globe

I no longer want to cower in the face of toxic systems
and my own self doubt

Though materially we need to live within
planetary boundaries

Our collective spiritual and relational limits are expansive

Untapped, waiting for more of us to find our rebel hearts

Mine currently teetering on the threshold

Has so much to learn

Small steps are steps still and we can take them quickly

Earth protectors spring up from every corner

The spirit perennial

Is there one essential substance and tone to a rebel

One beating heart

Or many percolating and half told stories
That grow from fragile seed
Tales of intention, duty and love
Bursting with potential and huge in capacity.

About the Author

Tom is a Poet living in Bristol, UK. His day job concerns play, nature and adventure for young people. This influences his poetry in the sense that he as interested in transformation. Tom has a poetry collection titled *Paint Yourself* published by Arkbound. He has poems in the anthologies produced by Proost titled *Learning to Love* and *Reaching for Mercy*. Tom is part of the Urban Word Collective and some of his poems feature in their anthologies titled *Lyrically Justified*.